Our Universe

Earth

by Margaret J. Goldstein

Lerner Publications Company • Minneapolis

Lerner Publications Company
A division of Lerner Publishing Group
241 First Avenue North
Minneapolis, MN 55401 USA

Website address: www.lernerbooks.com

Words in **bold** type are explained in a glossary on page 30.

Library of Congress Cataloging-in-Publication Data

Goldstein, Margaret J.
       Earth / by Margaret J. Goldstein.
          p.   cm. — (Our universe)
       Includes index.
       Summary: An introduction to the planet Earth, discussing
    its place in the solar system, movement in space, interior,
    surface features, atmosphere, weather, inhabitants, and
    moon.
       ISBN: 0–8225–4650–7 (lib. bdg. : alk. paper)
       1. Earth–Juvenile literature. [1. Earth.]
    I. Title. II. Series.
    QB631.4 .G65 2003
    525–dc21   2002002917

Manufactured in the United States of America
1 2 3 4 5 6 – JR – 08 07 06 05 04 03

The photographs in this book are reproduced with permission from: © Pat and Chuck Blackley, pp. 3, 19, 22, 27; NASA, pp. 4, 5, 9, 16, 17, 25, 26; © A.A.M. Van Der Heyden, p. 18; National Space Science Data Center, pp. 20, 21; © Wisconsin Department of Natural Resources, p. 23.

Cover: NASA.

This planet is home to plants, animals, and people. It is the place where you live. What planet is this?

This planet is Earth. Earth is our home
planet. It is part of a group of planets
called the **solar system.**

There are nine planets in the solar system including Earth. The Sun is a **star** at the center of the solar system. Earth is the third planet from the Sun.

All of the planets in the solar system **orbit** the Sun. To orbit the Sun means to travel around it.

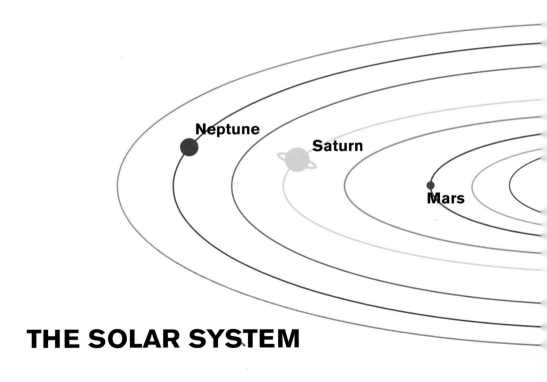

THE SOLAR SYSTEM

It takes Earth 365 days to orbit all the way around the Sun. We call this amount of time one year. How else does Earth move in space?

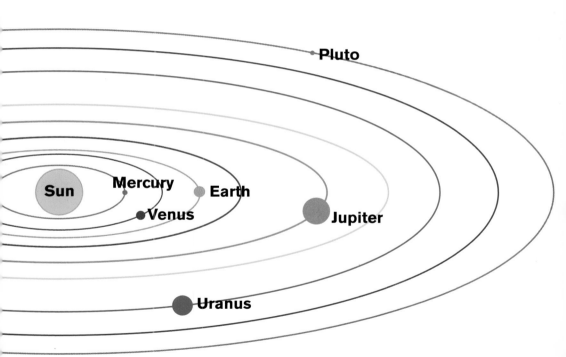

Earth also spins around like a top. This kind of spinning is called **rotating.** Earth rotates all the way around in 24 hours. We call this amount of time one day.

Earth rotates on its **axis.** An axis is an imaginary line running through the center of a planet. Earth's axis is tilted. That means Earth is tilted, too.

Rotating creates day and night. In the evening, your half of Earth rotates away from the Sun. The sky grows dark. It is night.

In the morning, your half of Earth rotates to face the Sun. Sunshine lights up the sky. It is daytime. Half of Earth is always dark when the other half is light.

DAY AND NIGHT

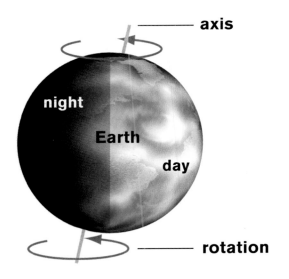

axis

night

Earth

day

rotation

Sun

SEASONS ON EARTH

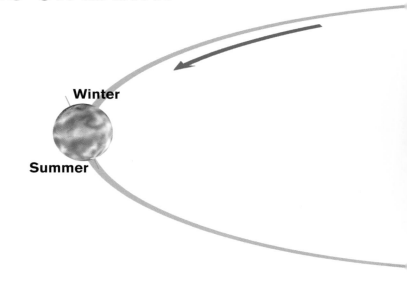

Earth's tilt gives our planet its **seasons.**
Your part of Earth tilts toward the hot Sun
for part of the year. The weather grows
warm. It is summer where you live.

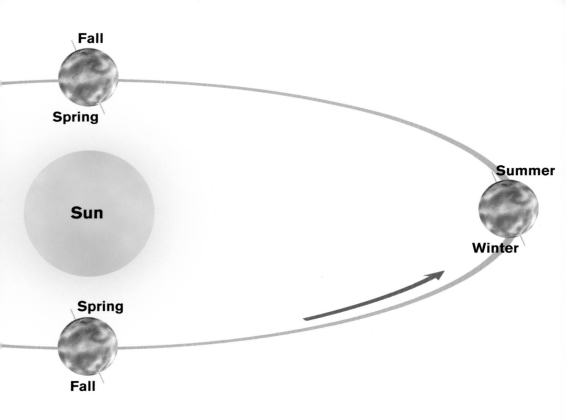

Later in the year, your part of Earth tilts away from the Sun. The weather turns colder. It is winter where you live. The seasons keep changing as Earth orbits the Sun.

Earth is a solid planet made of rock and metal. The center of Earth is called the **core.** It is a ball of hot metal. Above the core is a thick layer of rock. This layer is called the **mantle.** Some of the rock in the mantle is so hot that it is melted. A thin layer of hard rock covers the mantle. This outer layer is called the **crust.**

EARTH'S LAYERS

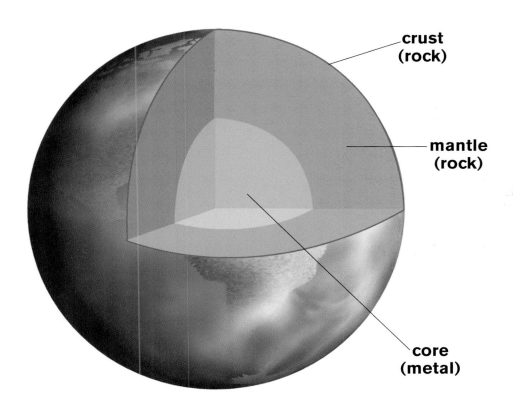

crust
(rock)

mantle
(rock)

core
(metal)

Water covers most of Earth's crust.
The oceans hold most of the water.
Our planet also has many lakes, rivers,
and streams.

The rest of the crust makes up the land on Earth. The largest areas of land are called **continents.** There are seven continents in all.

There are many different kinds of land on Earth. There are flat places and hilly places. There are tall mountains and deep valleys.

There are sandy deserts and green forests. Some land is covered with tall grasses. Other land is covered with ice and snow.

A layer of gases lies above the ground on Earth. This layer is called the **atmosphere.** The atmosphere is like a blanket around the planet. It helps to hold in heat from the Sun.

The atmosphere is where our weather begins. Clouds float in the atmosphere. Winds blow and storms rage. Rain and snow fall to the ground.

Trees and flowers live on Earth. People, birds, fish, and other animals live here, too. Earth is the only planet known to have living things. What makes Earth special?

Earth has everything plants and animals need to live. It has water to drink. It has air to breathe. And it is not too close to the Sun or too far from the Sun. So it is not too warm or too cold.

Earth has a nearby neighbor in space. This neighbor is the Moon. The Moon is a round ball of rock. It orbits Earth the way Earth orbits the Sun.

Astronauts are people who travel to outer space. Astronauts have studied Earth and the Moon. They have taken pictures of our planet from space. They have visited the Moon.

You do not have to be an astronaut to study Earth. Look at the sky, the land, and the water around you. Ask your own questions about your home planet. What would you like to know?

Facts about Earth

- Earth is 93,000,000 miles (150,000,000 km) from the Sun.

- Earth's diameter (distance across) is 7,930 miles (12,800 km).

- Earth orbits the Sun in 365 days.

- Earth rotates in 24 hours.

- The average temperature on Earth is 59°F (15°C).

- Earth's atmosphere is made of nitrogen and oxygen.

- Earth has one moon.

- Water covers about 70 percent of Earth's surface.

- About 50 bodies the size of the Moon could fit inside Earth.

- Earth is the fifth largest planet in the solar system.

- From space, Earth looks blue and white. The blue areas are oceans. The white areas are clouds.

- Earth is at least 4.5 billion years old.

- Earth orbits the Sun at about 66,700 miles (108,000 km) per hour. That's about 1,000 times faster than a moving car.

- In 1958, the United States launched a spacecraft from Earth for the first time. The spacecraft was called *Explorer 1*. *Explorer 1* orbited Earth and studied our planet from space.

Glossary

astronaut: a person who travels into space

atmosphere: a layer of gases that surrounds a planet or moon

axis: an imaginary line that goes through the center of a planet

continents: the seven large land areas on Earth

core: the center of a planet

crust: the outer layer of a rocky planet

mantle: a layer of rock that surrounds the core of a rocky planet

orbit: to travel around a larger body in space

rotating: spinning around in space

seasons: four periods of the year, which change as Earth travels around the Sun. The seasons are winter, spring, summer, and fall.

solar system: the Sun and the planets, moons, and other objects that travel around it

Learn More about Earth

Books

Brimner, Larry Dane. *Earth.* New York: Children's Press, 1999.

Murray, Peter. *Planet Earth.* Chanhassen, MN: Child's World, 1998.

Websites

Solar System Exploration: Earth
<http://solarsystem.nasa.gov/features/planets/earth/earth.html>
Detailed information from the National Aeronautics and Space Administration (NASA) about Earth, with good links to other helpful websites.

The Space Place
<http://spaceplace.jpl.nasa.gov>
An astronomy website for kids developed by NASA's Jet Propulsion Laboratory.

StarChild
<http://starchild.gsfc.nasa.gov/docs/StarChild/StarChild.html>
An online learning center for young astronomers, sponsored by NASA.

Index